T9425

DATE DUE

THE NEW CREEPY CRAWLY COLLECTION

FLEAS

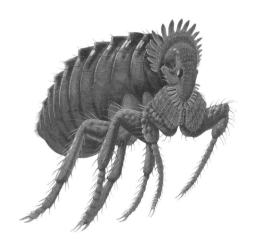

For a free color catalog describing Gareth Stevens' list of high-quality books and multimedia programs, call 1-800-542-2595 (USA) or 1-800-461-9120 (Canada). Gareth Stevens Publishing's Fax: (414) 225-0377. See our catalog, too, on the World Wide Web: http://gsinc.com

Library of Congress Cataloging-in-Publication Data

Fisher, Enid.
    Fleas / by Enid Broderick Fisher ; illustrated by Tony Gibbons.
      p. cm. -- (The New creepy crawly collection)
    Includes bibliographical references and index.
    Summary: Examines the anatomy, behavior, habitat, dangers, and uses of these tiny creatures.
    ISBN 0-8368-1913-6 (lib. bdg.)
    1. Fleas--Juvenile literature. [1. Fleas.] I. Gibbons, Tony, ill. II. Title. III. Series.
QL599.5.F57 1997
595.77'5--dc21   97-7341

This North American edition first published in 1997 by
Gareth Stevens Publishing
1555 North RiverCenter Drive, Suite 201
Milwaukee, Wisconsin 53212 USA

This U.S. edition © 1997 by Gareth Stevens, Inc. Created with original © 1996 by Quartz Editorial Services, 112 Station Road, Edgware HA8 7AQ U.K.

Consultant: Matthew Robertson, Senior Keeper, Bristol Zoo, Bristol, England.

Printed in Mexico

1 2 3 4 5 6 7 8 9 01 00 99 98 97

THE NEW
CREEPY CRAWLY
COLLECTION

FLEAS

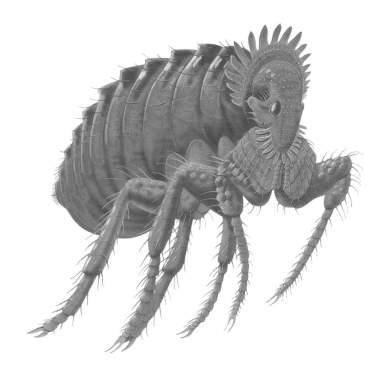

by Enid Broderick Fisher
Illustrated by Tony Gibbons

Gareth Stevens Publishing
MILWAUKEE

Contents

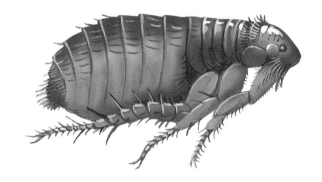

Getting to know fleas

If an animal starts scratching, as the rat *below* is doing, then chances are it has fleas.

Fleas (*right*) are nasty little creatures. They are only the size of specks of dust, but they can jump from 8 inches (20 centimeters) to an incredible 12 inches (30 cm) in a single leap. That's the equivalent of someone your size leaping nearly 1,150 feet (350 meters)!

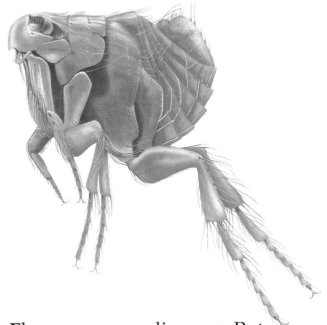

But jumping is not the only thing fleas can do well. They can also bite, and very painfully, too.

Fleas can carry diseases. Rat fleas, for example, once spread the plague in Europe. Humans have been trying to wipe the fleas out for centuries, but so far they have not been successful.

So what gives fleas their amazing jumping power? Why do they bite us? Is there anything about them we *can* like? Turn the pages that follow to discover all about these irritating little creatures.

Under the

Fleas are so small — less than 0.5 inch (1–10 millimeters) long — that you need to look at them under a microscope to study them. In fact, some are smaller than the period (.) at the end of this sentence. This flea has been enlarged about thirty-five times. Fleas are usually brown, which provides excellent camouflage on most animals.

Fleas have no respect for their victims. They will jump on to animals or humans, bite, suck their blood, and run away when a giant paw or hand tries to squash them. They don't have wings because they don't need to fly. And some don't even have eyes! Once they land on a living creature, known as a host, they simply feel their way along.

As you can see, a flea has three main body parts — head, thorax (chest), and abdomen. It also has six legs, and its body is so thin it looks as if it may have been ironed flat!

microscope

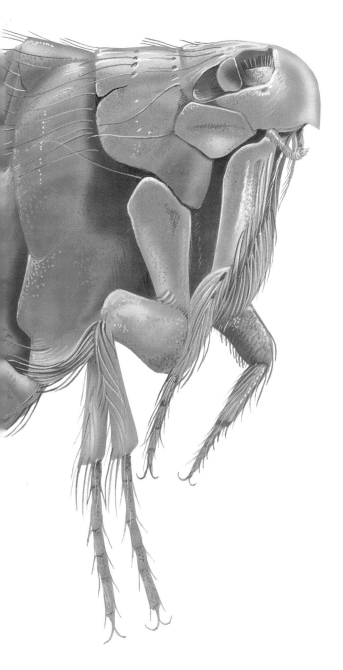

A flea has several spikes on its head. The two at the front are the antennae, which are used for touching or feeling. The longer, fearsome-looking spikes are the mouthparts. These are called stylets and have jagged, cutting edges, like a steak knife. The flea uses them to bite its host and suck up its blood. The flea shown here has eyes, but its eyesight is very poor.

The secret of the flea's jumping power is a hard pad made out of an elastic substance called resilin in its thorax. This pad is pulled back like a catapult when the flea bends its back legs. As the legs straighten out, the pad then straightens up and hurls the flea forward.

The flea's abdomen is covered in ten firm, overlapping plates. These will protect its soft insides if a creature scratches at it. A flea will survive attempts to squash it unless these plates are broken, when they will make a cracking sound.

The Black Death

Drinking a meal of blood might sound disgusting, but fleas can't live on anything else. If there are plenty of animals — or people — nearby, they usually eat once a day. But if they can't find anything suitable, they can go for months without sucking a single drop. When they eventually find a victim, however, they will have a feast.

Fleas use three razor-sharp points, called stylets, to punch holes in a victim's skin. Then they spit saliva down special tubes in the stylets and into the holes. At this point, they can begin to suck up some blood. With luck, they can get enough food to last the day with just one bite. But if their host scratches and interrupts the start of the meal, then the fleas will have to bite again.

Fleas can sometimes make people very ill. This happens if the fleas suck up germs from a diseased animal and inject them into a victim along with their saliva.

Ancestors of today's rat fleas actually wiped out twenty-five million people in Europe in the middle of the fourteenth century, in an outbreak of illness known as the Black Death. One-quarter of the population died!

The streets at that time were infested with rats. Their fleas would suck up bubonic plague germs from the rats and pass them on to humans. Fleas can still pass on diseases today, although people are generally cleaner and there are medicines now that cure many illnesses.

8

Birth of

New fleas come into the world as very tiny, pearly-white eggs. The female flea mates only once and will then lay about twenty-five eggs each day for the rest of her life. This is normally about three or four weeks, so a single flea may lay up

The larvae are long, like worms, and still white, but they have no eyes or legs. They do not suck blood, but live on dust and bits of dead skin and dried blood.

to a total of seven hundred eggs in her lifetime. You can see some of the eggs greatly enlarged *above.* They were laid on a cat's fur.

After a few days, or sometimes as long as two weeks, the eggs turn into larvae.

As they get older, they need fresh blood, which the mother flea gives them.

a flea

She does this by passing it out through her bottom — not a very nice thought!

Flea larvae often get shaken out of an animal's fur, but can survive equally well in a host's bedding. After two or three weeks, each larva turns into a pupa, which then weaves a cocoon around itself.

Fleas like warmth, and developing fleas will delay maturing in cold weather. Pupae that are laid as eggs in winter do not usually hatch until the first sign of spring.

Then they will start jumping and biting, like the enlarged ones shown here.

If nothing is done to kill them, the poor cat will be really irritated!

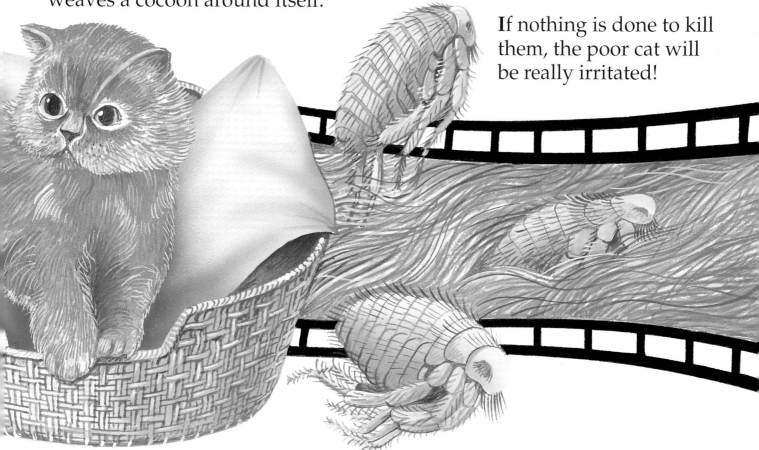

The flea's habitat

Animals such as badgers, foxes, squirrels, rats, mice, cats, dogs, and even humans — all enjoy a comfortable home. For this reason, they make perfect hosts for fleas.

Some animals are not good hosts. Monkeys, for instance, don't usually have fleas. And animals that spend a lot of time in water, such as otters, aren't much use, either. These animals do not have a suitable living area, or habitat, in which a flea's eggs can develop.

Most fleas rarely stay with the same animal, though. Badgers and foxes often have fleas; and humans can catch fleas from them, too. Cat fleas can easily jump on to a dog if there are no cats around, or on to you!

The female tropical jigger flea from South America waits for passing creatures to walk nearby. It then jumps on to the victim's feet and burrows into the flesh, usually just under a toenail. Ouch! There, it lays its eggs, which will hatch and eventually bite, causing ugly, painful sores.

Catching fleas

The Rothschild family is famous throughout the world for its work in banking, but one member of the family became famous for something much different. Miriam Rothschild was the first to discover how fleas jump.

Miriam grew up with a collection of fleas. Her father, Nathaniel, was an amateur collector, and he traveled the globe looking for new species.

He even met Miriam's mother while he was on a nature expedition to trap mice so that he could study their fleas!

He kept his fleas in a museum near the family home at Tring Park in Hertfordshire, England. By the time Miriam was born in 1908, the collection had thirty thousand specimens and was the best in the world. It was donated to the British Natural History Museum in London in 1913.

Miriam was fascinated by fleas and was the first scientist to discover the importance of resilin. This is the rubber-like substance that fleas rub their legs against and which gives fleas their amazing jumping power. She also studied how high and how far different types of fleas could jump.

There are books on many other creatures, such as butterflies, written by Miriam Rothschild, and she has written over a quarter of a million words on the subject of fleas alone! In fact, among her writings are the first five volumes of the British Natural History Museum's catalogue of fleas. Her scientific work was recognized by Oxford University, which awarded her a Doctor of Science degree. She was also made a Fellow of the Royal Society of Great Britain, one of the greatest honors a scientist can be awarded.

In the past, when people were not as hygienic as they are today, they had to find good ways of catching fleas — not in order to study them but simply to get rid of them. The underwiring of a lady's crinoline skirt, *right*, from the eighteenth century shows two china tubes hanging down from the waistband. These would be filled with a teaspoon of honey. People thought this would attract fleas that were nearby. The fleas would become stuck and be unable to bite. What an ingenious way of trying to catch them!

The flea

Step right up! We're taking a journey back in time to a sunny summer day one hundred years ago, and you are off to the fair. You'd better get there early, or you won't get a good view of the latest craze — the flea circus. You'll be able to see tiny fleas actually seem to race, pull tiny coaches, and perform all sorts of wonderful tricks. The circus master keeps them in an enclosed sand pit, but sometimes you can only see them when they disturb the dust. This is because the fleas he uses are, of course only a fraction of an inch (cm) long. However, we have enlarged them in this illustration to get a closer look.

Some flea-circus showmen had fleas that seemed to sword fight. Two fleas would face each other, with small swords tied to their legs by wire that was finer than your hair. As they waved their legs, the little swords clashed.

circus

The showmen would say that they spent hours training the fleas, but they were only teasing. You can't actually train a flea, but you *can* use its natural movements to create what seem to be miniature sporting events. The real skill was in building the props and then catching the fleas to perform. It must have been interesting to watch them running back and forth, pulling little chariots or performing on a tightrope!

Caring

for pets

Have you ever picked up your pet and, five minutes later, found yourself scratching like mad? You may have been bitten by a flea that your pet has picked up.

How did this happen? Your pet came close to a flea-ridden animal, and some of the fleas jumped on to your pet as it passed by. While a dog probably loves to chase and catch cats, mice, or birds, it can also get more than it bargained for if fleas decide to change hosts.

Once a pet is infested, you must act quickly, or soon your house will be jumping with fleas. Your pet's health is under threat, too. Scratching flea bites can cause ugly sores. These may fester and cause your pet a lot of discomfort.

Regular combing through a cat's or dog's coat will usually remove any fleas, as well as unwanted eggs, larvae, and pupae.

There are other things, too, you can do to get rid of pet fleas. Pet shops and some supermarkets also sell flea collars containing chemicals that kill fleas. These substances are very strong and can irritate the skin of some animals. Remove the collar at once if you see any signs of soreness around your pet's neck.

Once you know that fleas are around, you should also check your pet's bedding. This is a favorite breeding ground for fleas. Fleas like warmth and will also live in cozy corners, underneath radiators, or behind ovens.

Your parents can buy special chemicals to kill fleas imbedded in carpets, bed mattresses, and cushions. These chemicals are very poisonous, so never touch them yourself and do not allow your pet near them. Of course, a veterinarian can also give advice on what to do if a pet is infested with fleas.

Other

Fleas might be the world's worst pests, but many other creepy-crawly creatures can make our lives miserable, too!

Head lice

These tiny, brown parasites live in hair and suck blood from the scalp. They take about ten days to hatch, leaving behind empty, white egg cases called nits. You can catch lice if you wear hats or use combs that belong to someone who already has them or if your head comes too close to someone who has them. The lice can just walk off one head and on to another! You can get rid of them by wash-ing your hair with a special shampoo.

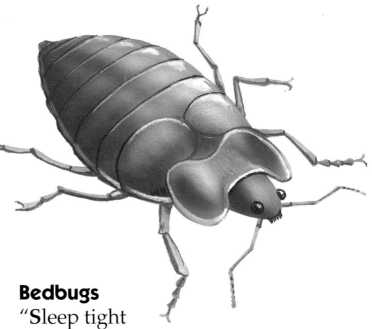

Bedbugs

"Sleep tight — don't let the bedbugs bite!" This humorous little bedtime rhyme is about bedbugs, tiny creatures like the one *above*, which is shown here highly enlarged. They hide in dirty mattresses and bedclothes. When they become active, they bite and suck the blood of the sleeper, causing itchy spots. They are greedy, too. One can suck up to seven times its own weight in just one meal. Just imagine eating your way through as much as seven times *your* weight for dinner!

pests

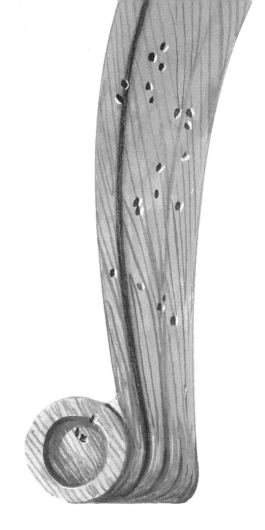

Book lice
These tiny creatures cannot read, of course, but they are still book-lovers and lay their eggs in damp, airless places. A little-used library is ideal. Each tiny louse (*plural* lice) lays its eggs between the pages of a book, and the hatching larvae live on any mold that has grown, damaging the paper as they eat away at it greedily.

Silverfish
About half the size of your little fingernail, these shiny, wingless creatures are not fish as their name suggests, but creepy-crawlies. They have bodies shaped like a miniature silver carrot, with long antennae and three tails that stick out behind. From this description, you should know one if you ever see one darting around quickly, as they tend to do. Silverfish often infest food cupboards; they feed on scraps of paper, spilled flour, and many other food-related items. They are pests in the kitchen.

Woodworms
A severe attack of woodworms can ruin good furniture, as you can see from the table leg *above*. Woodworms are the larvae of furniture, or death-watch, beetles. These hatch inside the wood and eat their way out, leaving tiny exit holes. Wood can be sprayed to kill the woodworms as they reach the surface, but the damage has usually been done by this time.

Did you know?

▼ What is the world's most popular host for fleas?
The red squirrel, *below*, is host to most fleas. Scientists have even found as many as thirteen thousand fleas on a single squirrel! Female hen stick-tight fleas also gather by the thousands on the heads and necks of poultry. They live there permanently, constantly annoying the poor chickens!

Can a flea bite make you ill?
Some flea bites are harmless. But if a flea has been sucking the blood of a diseased animal, and then bites you, it passes on that disease through the injected saliva. Some of these diseases are very dangerous. Prompt medical attention usually provides a cure, though.

Do fleas live alone or in colonies?
Most fleas lead independent lives, following supplies of food, but they may also live alongside several hundred others, if food is plentiful. Female hen stick-tight fleas (*Echidnophaga gallinacea*), however, will gather in thousands on the bare skin around the head and neck of poultry and live there permanently.

What sort of temperature suits fleas best?
Fleas breed best in warm, dry conditions, although those that inhabit Arctic regions have adapted to sub-zero temperatures.

How long do fleas live?
Adult fleas can live for quite a while, provided there is a steady supply of blood to suck. A human flea has even been known to live for nearly seventeen months.

What does a flea bite look like?
It first appears as a small, red lump caused by the flea injecting saliva into its victim. This causes itchiness; but if you get a flea bite, try not to scratch it.

How far can a flea jump?
Amazingly, fleas can jump from 100 to 200 times their own length, depending on the species.

▲ Which is the most dangerous flea in the world?
The oriental rat flea (*Xenopsylla cheopsis*) lives on rats that carry deadly diseases, such as the plague and typhus fever. It is highly likely to pass these on and is therefore very dangerous.

▼ How long have fleas existed?
Fossils of fleas found in Melbourne, Australia, are believed to be about 200 million years old. They were certainly around when dinosaurs, like the one shown here scratching itself, walked Earth long before humans existed.

Glossary

antennae — movable sensory organs, or feelers, on the head of an insect that are used for touching and smelling.

camouflage — markings or coloring that help an animal or plant disguise itself and blend in with its natural environment.

cocoon — the casing that an insect spins around itself and in which it develops into an adult.

host — an animal or plant upon which a parasite feeds.

larva — the wingless stage of an insect's life cycle between egg and pupa.

mate (v) — to join (animals) together to produce young.

parasites — organisms that live in or on other organisms.

plague — a highly contagious disease that can often cause death.

pupa — the stage of an insect's life between larva and adult.

species — animals or plants that are closely related and often similar in behavior or appearance. Members of the same species can breed together.

stylet — a sharp, pointed, piercing organ or body part.

thorax — an animal's middle section, or chest cavity, which holds the heart and lungs.

Books and Videos

Animals. *Under the Microscope* series. (Gareth Stevens)

Insect Attack. *Disaster!* series. Christopher Lampton (Millbrook Press)

Insect World. *Understanding Science and Nature* series. (Time-Life)

Insects. Joni P. Hunt (Silver Burdett)

The Insects. (Educational Images, Ltd. video)

Insects Harmful to Man. (International Film Bureau)

Index